UNDERSTANDING THE LIMITS OF ARTIFICIAL INTELLIGENCE FOR WARFIGHTERS

VOLUME 5_ MISSION PLANNING

KELLER SCHOLL

GARY J. BRIGGS

LI ANG ZHANG

JOHN L. SALMON

PREPARED FOR THE DEPARTMENT OF THE AIR FORCE

APPROVED FOR PUBLIC RELEASE; DISTRIBUTION IS UNLIMITED.

RAND PROJECT AIR FORCE

05

For more information on this publication, visit **www.rand.org/t/RRA1722-5**.

About RAND

RAND is a research organization that develops solutions to public policy challenges to help make communities throughout the world safer and more secure, healthier and more prosperous. RAND is nonprofit, nonpartisan, and committed to the public interest. To learn more about RAND, visit www.rand.org.

Research Integrity

Our mission to help improve policy and decisionmaking through research and analysis is enabled through our core values of quality and objectivity and our unwavering commitment to the highest level of integrity and ethical behavior. To help ensure our research and analysis are rigorous, objective, and nonpartisan, we subject our research publications to a robust and exacting quality-assurance process; avoid both the appearance and reality of financial and other conflicts of interest through staff training, project screening, and a policy of mandatory disclosure; and pursue transparency in our research engagements through our commitment to the open publication of our research findings and recommendations, disclosure of the source of funding of published research, and policies to ensure intellectual independence. For more information, visit www.rand.org/about/research-integrity.

RAND's publications do not necessarily reflect the opinions of its research clients and sponsors.

About This Report

This is the fifth report in a five-volume series addressing how artificial intelligence (AI) could be employed to assist warfighters in four distinct areas: cybersecurity, predictive maintenance, wargames, and mission planning. These areas were chosen to reflect the wide variety of potential uses and also to highlight different kinds of limits to AI application. Each use case is presented in a separate volume, as it will be of interest to a different community.

This fifth volume provides a description of how AI can be used to conduct mission planning and how these methods compare with more traditional operations research approaches. It is aimed at those with an interest in mission planning, operations research, and AI applications more generally. Volume 1 in the series provides a summary of the findings and recommendations from all use cases, and the other volumes provide detailed analysis of the individual use cases:

- Lance Menthe, Li Ang Zhang, Edward Geist, Joshua Steier, Aaron B. Frank, Eric Van Hegewald, Gary J. Briggs, Keller Scholl, Yusuf Ashpari, and Anthony Jacques, *Understanding the Limits of Artificial Intelligence for Warfighters*: Vol. 1, *Summary*, RR-A1722-1, 2024
- Joshua Steier, Erik Van Hegewald, Anthony Jacques, Gavin S. Hartnett, and Lance Menthe, *Understanding the Limits of Artificial Intelligence for Warfighters*: Vol. 2, *Distributional Shift in Cybersecurity Datasets*, RR-A1722-2, 2024
- Li Ang Zhang, Yusuf Ashpari, and Anthony Jacques, *Understanding the Limits of Artificial Intelligence for Warfighters*: Vol. 3, *Predictive Maintenance*, RR-A1722-3, 2024
- Edward Geist, Aaron B. Frank, and Lance Menthe, *Understanding the Limits of Artificial Intelligence for Warfighters*: Vol. 4, *Wargames*, RR-A1722-4, 2024.

The research reported here was commissioned by Air Force Materiel Command, Strategic Plans, Programs, Requirements and Assessments (AFMC/A5/8/9) and conducted within the Force Modernization and Employment Program of RAND Project AIR FORCE as part of a fiscal year 2022 project, "Understanding the Bounds of Artificial Intelligence in Warfare Applications."

RAND Project AIR FORCE

RAND Project AIR FORCE (PAF), a division of the RAND Corporation, is the Department of the Air Force's (DAF's) federally funded research and development center for studies and analyses, supporting both the United States Air Force and the United States Space Force. PAF provides the DAF with independent analyses of policy alternatives affecting the development, employment, combat readiness, and support of current and future air, space, and cyber forces. Research is conducted in four programs: Strategy and Doctrine; Force Modernization and Employment; Resource Management; and Workforce, Development, and Health. The research reported here was prepared under contract FA7014-22-D0001.

Additional information about PAF is available on our website:
www.rand.org/paf/

Acknowledgments

We thank our sponsor contact, Kathryn Sowers, and our action officers, Julia Phillips and
Gregory Cazzell, for their guidance in choosing the use cases, for their thoughtfulness in scoping the
research questions, and for working diligently with us to obtain the data necessary to conduct the
many machine-learning experiments described in this series of reports. Thanks as well to the following
individuals: R. Scott Erwin and Jean-Charles Ledé for graciously connecting us with many AI
development efforts across the Air Force Research Laboratory, Lee Seversky for sharing his expertise
on mission planning, Lt Col Kari K. Mott for discussing automation in Air Operation Centers, and Lt
Col Nicholas J. Harris for speaking with us across a dozen time zones to explain the Master Air
Attack Planning process.

We are also grateful to many current and former RAND colleagues, including Caolionn
O'Connell, Sherrill Lingel, Osonde Osoba, and Chris Pernin for helping us shape the research agenda.
We appreciate the support of our reviewers, Nick O'Donoughue and Jair Aguirre, without whom this
would be a weaker and less clear report, and the support we received from the entire Project AIR
FORCE team at RAND. We could not have written these reports without their help; any errors that
remain are ours alone.

Summary

Issue

Mission planning involves the assignment of discrete assets to prioritized targets, including the dynamic routing of those assets to their destinations under complex environmental conditions. Because of the value of quick turnaround and the relative simplicity of the simulated operational environment, there has been considerable interest in improving the mission planning process with the addition of reinforcement learning techniques for artificial intelligence (AI), which could produce better, faster, or simply unique solutions for human consideration.

Approach

For background research, we interviewed subject-matter experts, examined academic literature and government and industry publications, and investigated a compelling state-of-the-art model trained on a similar domain (AlphaStar).[1] As a result of the lessons learned from AlphaStar, we embarked on two lines of original research that focused on applying AI to narrow route-planning missions. We experimented with in-house modeling software to quantify and better understand the relative strengths and weaknesses of AI models in general. We also implemented a solution to a substantial barrier (the lack of systems integration) to understand that barrier more precisely and what could be done to overcome it.

Key Findings

It is very difficult to predict how fast AI will advance. Even so, there are some areas in which AI will be relatively capable and for which building capacity, experience, and user trust are important steps in its further adoption. One example is the area of mission planning.

The current mission planning process is human intensive but sufficient for a peacetime tempo: No subject-matter expert or practitioner thought that the demands exceeded the capacity for planning. Although mission planning at the scale of an entire theater's assets is relatively rare, efforts to build capacity to automate the entire process would be beyond current AI capabilities and would struggle from severe data limitations.[2] On the basis of our subject-matter expert interviews and a case study, AI

---

[1] The interviews were conducted from September 2021 to May 2022 and took place primarily by phone or video call. The interviews were not for attribution, so no names are provided.

[2] The mission planning process includes the commander's intent; the selection of targets; the evaluation of threats and environmental considerations, such as inclement weather; the selection of air assets to observe points of interest and to neutralize targets; the mapping of how U.S. and allied air assets will accomplish their missions; the processing of all information generated

will be relatively weak at rapidly adapting to new enemy strategies and tactics. However, there are specific applications, such as rapidly responding to new information (e.g., unknown threats), for which AI can provide a substantial advantage.

Key issues in the adoption of AI are likely to be software integration and data availability, as those issues already impose substantial constraints on existing human processes. We validated that this assessment holds: Demand was strong for our solution to a software integration issue to link two popular software systems. We found that AI, even when it cannot provide optimal solutions, has a substantial speed advantage over alternatives and can tolerate changing conditions.

Table S.1 shows the difference between a mathematical approach that is commonly incorporated in operations research and our AI approach.

Table S.1. Comparison Between Operations Research and Artificial Intelligence Approaches for Route Planning

	Operations Research	AI
Enemy detection	Average risk level: 1.12	Average risk level: 2.91
Path length	Average: 5,070 km	Average: 3,938 km
Cost to develop	Low	High
Response time	Seconds to minutes	Milliseconds
Predictability	Fixed given a utility function and the situation	Varies with each run

The average risk of enemy detection was substantially lower for the traditional (operations research) method; the cost of development was lower, but it was more predictable, took longer to generate a response after new information, and charted longer paths.

Recommendations

We have three primary recommendations for the Department of the Air Force's AI work for mission planning:

- AI implementation for mission planning should target areas of relative human weakness, particularly reacting quickly to new situations, and plan to expand from there
- The DAF should prioritize creating not just useful tools and software but enabling those resources to be extended by others (both government and third parties) and connecting them to existing systems
- The DAF should continually monitor the AI landscape: paradigm shifts have happened before in AI and likely will again.

---

by those mission flights; and all the information flows and iteration required to make the mission function (since the availability of assets can limit which targets can be addressed on a given day).

Contents

Figures and Tables

Figures

Tables

viii

Chapter 1

Introduction

Overview

The Department of the Air Force has become increasingly interested in the potential for artificial intelligence (AI) to revolutionize different aspects of warfighting.[3] For this project, the U.S. Air Force asked RAND Corporation researchers to consider broadly what AI can and *cannot* do in order to understand the limits of AI for warfighting applications. To address this request, we investigated the applicability of AI to four specific warfighting applications: *cybersecurity, predictive maintenance, wargames,* and *mission planning.*

On the basis of a given Joint Integrated Prioritized Target List, planners in the Air Operations Center (AOC) allocate weapons and platforms to achieve a commander's objectives through the Master Air Attack Plan process. From these pairings, more-detailed mission route planning takes place by mission planners at the squadrons.

Most time spent on the mission planning process goes to gathering data from different agencies, translating the data, entering the data into software, and generating frequent PowerPoint briefings. The information is not managed well: Selecting a given strike package or asset does not prepopulate forms with relevant information, necessitating both manual entry and careful checks to ensure there are no errors, since incorrect values for operational ranges or effective radar distance could lead to disaster.

In a highly contested environment against a peer adversary, the current air tasking cycle of 72 hours is insufficient for air operations.[4] The Air Force Life Cycle Management Center is working on decreasing this time within the Joint Mission Panning System, but such efforts have been made before.[5] By creating faster and more-effective planning software, we might be able to speed up the planning process or make better plans within existing time constraints. Identifying which parts of the mission planning process can be augmented by AI, if any, is the focus of this report.

To examine how AI for mission planning could be used effectively, we looked at similar systems. Over the past decade, there have been several high-profile demonstrations of AI systems achieving high-level performance in games that were once thought to require human intelligence. Many of these demonstrations involved turn-taking games that take place on a fixed board, with complete

---

[3] In this report, we use *AI* to refer to the most common methods of machine learning (ML) today: deep learning neural networks.

[4] Sherrill Lingel, Edward Geist, Thomas Hamilton, Daniel M. Norton, and Colby P. Steiner, *(U) Toward Continuous Planning for Modern Warfare: A Warfighting-Focused Framework for Operational Planning of Science and Technology Pursuits,* RAND Corporation, RR-A953-1, 2023, Not available to the general public.

[5] K. Houston Waters, "Hanscom AFB Team Supports JADC2 Through Agile Software Development," U.S. Air Force, June 7, 2021; David E. Thaler and David A. Shlapak, "Perspectives on Theater Air Campaign Planning," RAND Corporation, MR-515-AF, 1995.

information available to both players. For example, DeepMind's AlphaZero demonstrated grandmaster-surpassing abilities in chess, go, and shogi.[6] The more recently developed AlphaStar is an AI system capable of outperforming expert humans in the StarCraft II video game.[7] AlphaStar was also created by DeepMind, a London-based AI startup that is now owned by Alphabet (formerly Google). DeepMind started working on mastering StarCraft II after finishing its conquest of turn-based games with the success of AlphaGo against Lee Sedol, a former professional go player, in 2016.[8]

StarCraft II is arguably one of the most complex and powerful examples of reinforcement learning (RL) capability within the field of AI and embodies many characteristics of operational and tactical command and control, including critical decisionmaking when faced with the unknown (the fog of war). We assessed the lessons learned from the development of AlphaStar to guide which aspect of mission planning is suitable for AI. In short, AlphaStar's strengths demonstrate that small tactical-level problems might be best-suited to AI, which led us to explore AI for route planning.

Game-Playing Artificial Intelligence: AlphaStar

AlphaStar's ability to play StarCraft II was used as a basis for our assessment of AlphaStar's AI strengths. StarCraft II is an intense, real-time strategy game with a highly competitive international scene. StarCraft II's most popular two-player mode follows three rough phases of gameplay. Players start with a home base and worker units that acquire resources. In the first phase, players begin constructing buildings and units; the players can send units to harass or simply scout the enemy to understand what they are doing. Players will choose upgrades and new options for acquisitions to counter their expectations of the enemy's choices. In the second phase (referred to as mid-game), forces will clash and players will expand to additional bases. In the last phase (endgame), upgrades are de-emphasized in favor of churning out units to win decisive fights.

StarCraft II presents a significantly harder AI problem than previous games. At a given time step, there are 10^{26} options on average, the equivalent of around ten moves by each player in chess. Each time step in StarCraft II happens in a small fraction of a second. The number of possible options at a microscale is immense, dwarfing such turn-based games as chess or go. See the Appendix for a more detailed discussion.

AlphaStar was far superior to humans at tactical decisionmaking and fast reactions, and it used its multitasking, tactical-level superiority to compensate for a lack of strategic understanding. AlphaStar did not rely on superior strategy to defeat humans because it could leverage its overmatch in battles to

---

[6] David Silver, Thomas Hubert, Julian Schrittwieser, Ioannis Antonoglou, Matthew Lai, Arthur Guez, Marc Lanctot, Laurent Sifre, Dharshan Kumaran, Thore Graepel, Timothy Lillicrap, Karen Simonyan, and Demis Hassabis, "A General Reinforcement Learning Algorithm That Masters Chess, Shogi, and Go Through Self-Play," *Science*, Vol. 362, No. 6419, December 2018.

[7] Vinyals, Oriol, Igor Babuschkin, Wojciech M. Czarnecki, Michaël Mathieu, Andrew Dudzik, Junyoung Chung, David H. Choi, Richard Powell, Timo Ewalds, Petko Georgiev, Junhyuk Oh, Dan Horgan, Manuel Kroiss, Ivo Danihelka, Aja Huang, Laurent Sifre, Trevor Cai, John P. Agapiou, Max Jaderberg, Alexander S. Vezhnevets, Rémi Leblond, Tobias Pohlen, Valentin Dalibard, David Budden, Yury Sulsky, James Molloy, Tom L. Paine, Caglar Gulcehre, Ziyu Wang, Tobias Pfaff, Yuhuai Wu, Roman Ring, Dani Yogatama, Dario Wünsch, Katrina McKinney, Oliver Smith, Tom Schaul, Timothy Lillicrap, Koray Kavukcuoglu, Demis Hassabis, Chris Apps, and David Silver, "Grandmaster Level in StarCraft II Using Multi-Agent Reinforcement Learning," *Nature*, Vol. 575, November 2019.

[8] Silver et al., 2018.

win most games. Later iterations showed improvements, but AlphaStar still struggled to adapt to new strategies.

In StarCraft II, players virtually divide gameplay into micro and macro elements. *Micro* refers to the command and control of individual units and is the predominant gameplay during battles. Multitasking plays a massive role in micro. *Macro* refers to strategic elements: determining army composition, developing research, and increasing economic output. AlphaStar won primarily using micro elements because machines have superior multitasking ability.

AlphaStar was able to win most battles by relying on a single-ranged unit even though in StarCraft II, this strategy typically leads to defeat because it is designed in such a way that units counter each another (e.g., similar to the hand game rock paper scissors).[9] Superior microstrategy enabled AlphaStar to win battles against armies with optimal countermeasures. Specifically, it took advantage of a ranged stalker unit with a short-range teleport and recharging shields that, under superhuman control, could teleport away at the precise moment before death, recharge its shields, and get back into the fight. The overmatch of AlphaStar in multitasking and battle tactics was sufficient to defeat professional human players.

Despite this overmatch, AlphaStar was comparatively weak across other areas in the game. Some early weaknesses include a lack of scouting (to determine army composition to counter the enemy), vulnerability to early attacks (a lack of defenses to counter enemy harassment on economic output), and the inability to create bottlenecks on the battlefield (placing buildings at anticipated chokepoints for an advantage during battles). AlphaStar fixed these weaknesses in later games but still had much more trouble with novel strategies, even after continued improvements.[10] This shows that AlphaStar struggled with strategic-level decisionmaking.

RL is designed to help an algorithm determine an action to receive a reward. Within StarCraft II's micro, battles are fought and won quickly, and it is easy to attribute rewards to a given action.[11] However, attributing rewards for a battle that was won because of the pixel-perfect placement of a chokepoint 10,000 time steps ago is a much more difficult problem. This problem of sparse rewards is an ongoing research area in the field and is the most likely reason why AlphaStar had difficulty competing on the strategic (macro) scale.

From the AlphaStar example, we identified two key lessons:

- We should expect fast reactions to be relatively better and strategic decisionmaking under uncertainty to be relatively worse compared with a human baseline. Strategic and operational planning capabilities will lag fast reactions and combat capabilities.

---

[9] In a more detailed example, imagine a scenario in which spearmen counter a cavalry, which counters archers, who counter spearmen, and AlphaStar wins with the equivalent of an all-archer army. A cavalry, which otherwise counters archers, is defeated because archers are world champion–level snipers who pick optimal targets at the individual level.

[10] For example, see LowkoTV, "StarCraft 2: AlphaStar (Artificial Intelligence) vs Grand Master League!" video, YouTube, November 12, 2019, 18:50–19:23.

[11] For example, AlphaStar learned that retreating while firing on the enemy is a good strategy because it keeps enemy units within firing range but its own units out of the enemy's range. The immediate reward is that the enemy units' health decreases while its own units' health stays high.

- There is no replacement for operational understanding and feedback loops. AlphaStar's designers included grandmaster StarCraft players and pure AI experts who actively intervened in the training process to correct errors.

Approach

AlphaStar demonstrated that AI (and RL in particular) can excel at narrow problems. Route planning is an area in which there has been AI success in the civilian sector, with self-driving cars and autopilots the most prominent examples.[12] Proper route planning can minimize risk to pilots and systems, reduce enemy information about U.S. assets, and increase the likelihood of successful mission execution.

Although only a subset of all route planning, planning for an individual package to penetrate enemy airspace contains enough elements for adequate complexity (enemy positions and multiple viable routes) without overloading either the reader or ourselves (with dynamic enemy positions, for example). It is also a scenario that is frequently encountered by the Department of the Air Force (DAF), and, thus, potentially worthwhile to implement because AI models have relatively higher up-front costs than training humans to do a task, but the AI models scale to large numbers more easily. Using a prior RAND internal model, we explore the feasibility of applying AI to this application, compare AI performance against an optimization approach, and assess the limitations of this approach. In this report, we compare a pure optimization approach to mission planning with a pure RL approach. Hybrid solutions that involve both approaches or include an experienced human in the process might yield improved outcomes.

Organization of this Report

Chapter 2 presents RAND researchers' work to identify relative and absolute weaknesses of ML models compared with computational alternatives. Chapter 3 describes our work to make an RL environment in the Advanced Framework for Simulation, Integration, and Modeling (AFSIM). This overcame a key barrier, integrating existing systems with existing AI toolkits. Chapter 4 contains our findings and recommendations.

---

[12] See Laura Fraade-Blanar and Brian A. Jackson, "Developing a Winning Safety Strategy for Automated Vehicles," *RAND Blog*, February 18, 2022.

Chapter 2

RAND Target Accessibility Model Investigations

We used a proprietary model called the RAND Target Accessibility Model (RTAM) to examine an ML model trained to penetrate enemy air defenses and reach a target point compared against an operations research (OR) optimization function designed to achieve the same goal. We were interested in the weak points of the ML approach on both an absolute and a relative basis. Given the pace of modern ML improvements, the lessons of history, and even the positive returns to scale on very large and expensive state-of-the-art models, we are hesitant to describe any current limitation as forever impossible.[13]

However, by examining *relative* strengths and weaknesses in a toy system, it should be possible to determine what tasks AI will be relatively better or worse at performing. This is a somewhat safe extrapolation as long as the current paradigm in ML, neural networks, continues to lead. The end of the current paradigm will be highly visible, making this a safe assumption to use, so long as it seems to continue to hold.

In our analysis, the OR approach found the better path than the ML model almost every time. We also found that ML models take substantially longer to train. This is not simply a matter of computer time: The longer training effort for a model means that poorly specified utility functions take longer to resolve. However, we see an advantage for ML in adaptability: It can make decisions about what to do given new information dramatically faster. This is particularly valuable under conditions of high uncertainty: Encountering enemy assets is more likely because they cannot be avoided but also more informative because encountering an enemy asset provides substantially more information about its location.

Modeling Approach

RTAM was developed by RAND over the last 20 years to enable like-for-like analysis of alternatives and to broadly explore how successfully different aircraft options can penetrate different air defense threat areas. It provides a simulated environment that is suitable for, among other tasks, directing a flying agent through an area populated by threats. The agent can be controlled manually with a controller or by any programmatically defined input. The size of the area within which the agent can be targeted can be manually adjusted, and the system supports both prepopulated known threats and pop-up threats that can be dynamically added to the simulation. We used an existing

---

[13] Jared Kaplan, Sam McCandlish, Tom Henighan, Tom B. Brown, Benjamin Chess, Rewon Child, Scott Gray, Alec Radford, Jeffrey Wu, and Dario Amodei, "Scaling Laws for Neural Language Models," arXiv, January 23, 2020.

RTAM model of one drone attempting to navigate the now-defunct Semi-Automatic Ground Environment (SAGE) early warning radar network across the continental United States (CONUS), avoiding detection to the extent possible and navigating to a specific mission objective in the interior. For simplicity, clarity, and training speed, we examined the traversal from a single starting point to a single end point: We did not expect our qualitative results to change when adding additional targets and determining the order of those targets.

The OR approach divides the continuous universe (see Figure 2.1) into discrete spaces with edges between nodes to span the area of interest. Given a specified utility function (primarily to minimize enemy radar detections), the OR goal is to find the mathematically optimal path within the possible edges. Risk at each edge is calculated, and a path of lowest risk is determined via a Dijkstra-like algorithm. A trade-off exists when geographically precise models use more-discrete spaces and, therefore, more time is required to calculate a path.

Figure 2.1. RAND Target Accessibility Model Map Showing Node-Edge Discretization

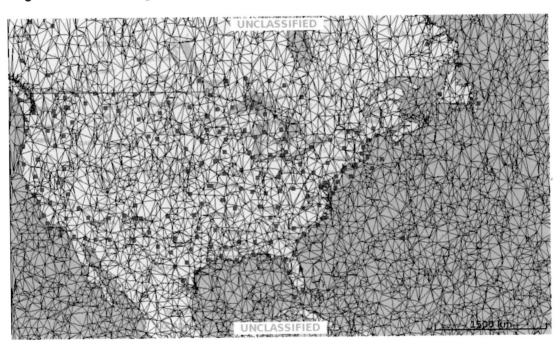

NOTE: A screenshot depicting an RTAM scenario. The lines represent possible traversal paths, after discretizing the environment, for the OR algorithm to choose. The small squares scattered around the United States represent early warning radars.

The ML approach uses RL. We used the Stable Baselines3 RL PyTorch neural network library and trained the model using the Proximal Policy Optimization (PPO) algorithm.[14] Training RL algorithms successfully is quite challenging, and, after much experimentation, we arrived at the following training heuristic. Leveraging the concept of curriculum learning, we first trained for a simple reach-the-target behavior, ignoring early warning radar detections. After the algorithm

---

[14] John Schulman, Filip Wolski, Prafulla Dhariwal, Alec Radford, and Oleg Klimov, "Proximal Policy Optimization Algorithms," arXiv, August 28, 2017.

generated reliable target-finding RL agents (any target from any starting position), we trained for early warning radar avoidance behavior using successive training sessions, starting with the base reach-the-target model.[15] The reward function, which will be discussed, was adjusted to incentivize desired behaviors. Model testing and evaluation was performed after each sequence of follow-on training explorations.

While the loss function (for OR) and the reward function (for RL) are not directly comparable because of algorithmic differences, both functions were designed to find the shortest path to a target while minimizing the total time detected by radars. The RL reward function required more terms, such as a reward for getting close to a target (and subsequently reaching the target to help motivate the agent during early training iterations), a penalty for turning, and a penalty for fuel (to encourage shorter paths).

Figure 2.2 is an image taken from RTAM part way through training an ML model.

Figure 2.2. Partially Trained Machine Learning Model Paths in the RAND Target Accessibility Model

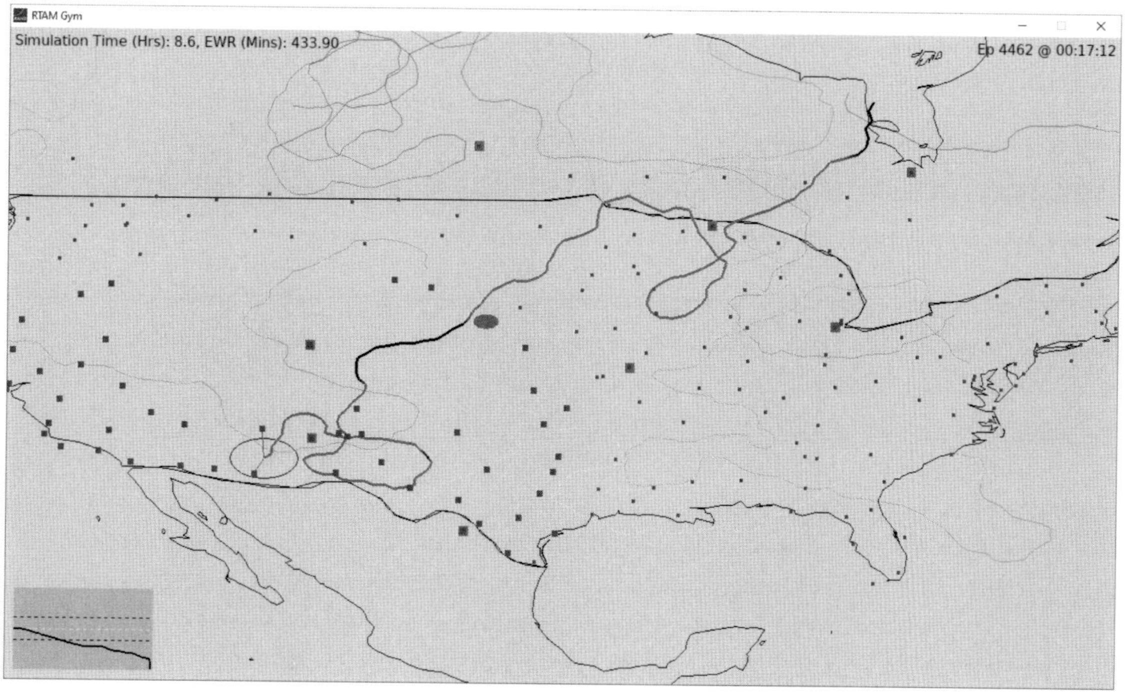

NOTE: An RTAM screenshot depicting ML-generated paths. The blue points represent the locations of early warning radar sites. The path segments are colored in red (detected by radar) or black (undetected). The circle at the end of the path is the target.

The red segments of the flight path indicate that the asset is being detected by a radar site and at risk. The black segments are sufficiently far from radar sites and not threatened. The faintly traced

---

[15] We used one million training epochs per session. This value was arbitrary and not rigorously tested, but we found it sufficient to reliably learn desired behaviors.

lines are paths from the most recent ten scenarios the agent was trained on, with a different starting position and ending target location.

The path that is shown is as inefficient as it looks: To make a model competitive with an OR approach requires a great deal of work to specify a utility function and environment, followed by thousands of training cycles. This process is never perfect, so many iterations of modifying the utility function or the training style are required. Each iteration will take hours compared with the seconds for an OR formula, leading to much slower observe-orient-decide-act loops. This lowers the quality of the resulting model and increases the cost dramatically. This disadvantage can be ameliorated with powerful servers and experienced engineers but not eliminated. Training only need be done once, offline, before the trained agent can be used operationally many times. But, in RTAM, the training is sensitive to aircraft signature, threat laydown, density, and types; this makes a true general-purpose route planning RL challenging but a narrowly focused one more viable.

RTAM OR provides a single analytic solution in the form of an *expected value* of detection time and shot opportunities; there is no variance in the expected value from run to run. On the other hand, the ML approach would vary significantly: A trained RL can be stochastic in nature, so for our analysis, we gave it ten attempts at each goal and chose the best one. When the time taken for the RL to calculate a path is milliseconds, a great deal of opportunity for finding optima exists depending on real-world timelines.

Quantifying Operations Research Versus Machine-Learning Performance: Experienced Risk

We employed several methods to describe the differences between the two methods. One visualization, Figure 2.3 and Figure 2.4, shows the summed time in enemy detection radius experienced by the OR approach (Figure 2.3) and the summed time in enemy detection radius by the ML model (Figure 2.4) as the aircraft traveled from a starting location in the middle of the Atlantic Ocean at a latitude and longitude of 34.86 and -61.85 degrees, respectively. These starting locations are indicated by a black diamond in Figures 2.3 and 2.4. The other points represent target locations spread throughout CONUS and reached via a path with a different risk level as the agent crosses through, or around, early warning radar threats. In most cases, the OR method produces a lower risk option (from this particular starting location) to each target between the ranges of 0 and 2.76 (note the scale of risk level to the right of Figure 2.3). In comparison, the risk level for the ML model extends as high as 8.18 for targets on the far west side of CONUS.

Figure 2.3. Risk Level During Ingress to Target for OR Paths

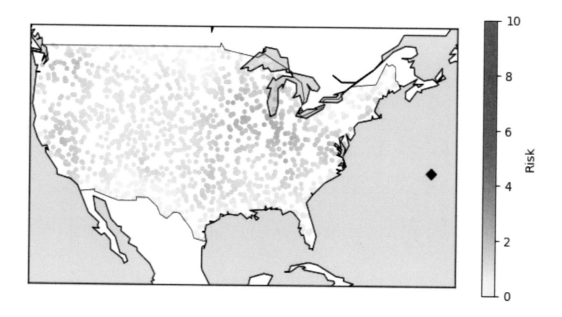

NOTE: The black diamond on the right is the starting location.

Figure 2.4. Risk Level During Ingress to Target for ML Paths

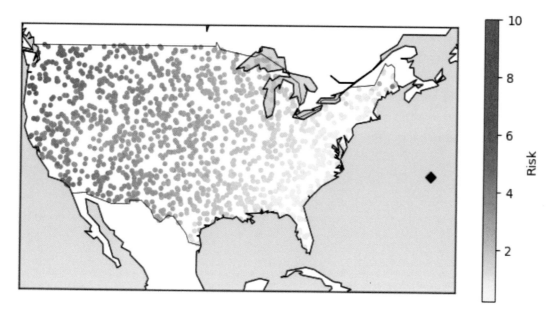

NOTE: The black diamond on the right is the starting location.

Figure 2.5 effectively combines Figures 2.3 and 2.4.

Figure 2.5. RAND Target Accessibility Model Image, Operations Research, and Machine Learning Advantage Map

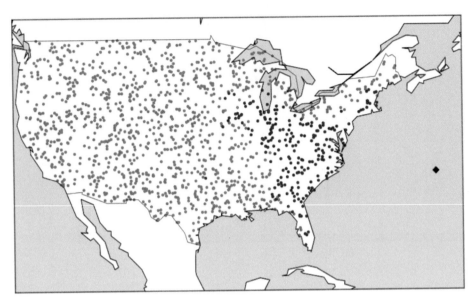

NOTE: The blue dots indicate ML outperformance, and the orange dots indicate OR outperformance. The black diamond on the right is the starting location.

The blue target locations indicate where the ML model outperforms the OR model, shown in orange, in terms of risk level. ML does better in only 16 percent of the tested locations, with a risk level that is on average lower than OR levels by 0.5. The ML model tends to outperform at approaching targets that are closer and on a more direct path to the starting point.

The same data are presented in two other ways to further illustrate the limitations of ML models in this domain. Figure 2.6 presents a histogram of all target risk levels. Risk levels for the ML model are presented in blue, and risk levels for the OR model are in orange.

Figure 2.6. Histogram of All Target Risk Levels

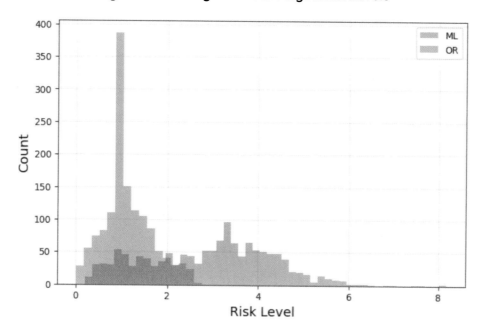

The overlaid histograms in Figure 2.6 of the risk levels for all 1,376 target locations for both models show a stark difference. As expected, most OR model risk levels are lower than the ML levels. The 16 percent of targets for which ML performed better are found within the overlaid portion of the two histograms. The ML model also has a much flatter distribution, with relatively more unusually risky and unusually safe paths. The very high peak of the OR model represents a more stable approach.

Finally, Figure 2.7 shows how the two risk levels compare with each other for a given target using a scatterplot format.

Figure 2.7. Operations Research Versus Machine Learning Outcomes Plot

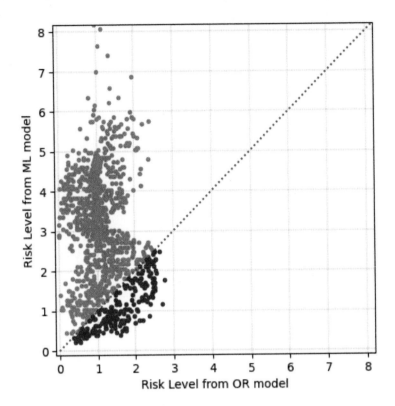

NOTE: A scatterplot comparing the risk levels for a given target (the scatter of dots along the x-axis represents the risk of the OR-generated path, and the dots along the y-axis represent the risk of the ML-generated path). The red dotted diagonal line is a $y = x$ plot indicating parity between the models. The orange scatter dots indicate when OR did better for a given target and the blue scatter dots when ML did better for a given target.

The target locations in blue correspond with the points where ML does better and is appropriately found below the 45-degree line of equivalence. The flatness of the distributions in Figure 2.7 is now the much greater vertical than horizontal spread along the axes. Interestingly, there is no particularly strong relationship between the risk levels of the two models: Whether the OR approach can find a good path seems to have little relationship to the ability of the ML model to do the same.

Path Length

The other key varying component of the utility functions is the length of the path the agent would use to reach the target. Shorter paths take less time, use less fuel, and cause less wear and tear. The ML model that we trained tended to take much straighter paths, going directly toward its goal with moderate adjustments to avoid enemy detection. In the simulation environment and with the agent we used, speed was constant, so the path length and total time flying were equivalent.

Figure 2.8 shows sample paths for three ML cases (labeled AI-1 to AI-3) and three OR cases (labeled OR-1 to OR-3).

Figure 2.8. Comparison of ML Paths to Corresponding OR Paths for Similar Target Locations

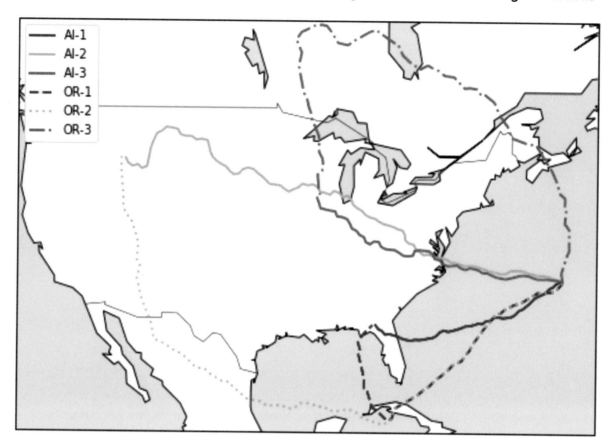

NOTE: The ML paths are indicated by the solid lines labeled AI-1, AI-2, and AI-3 in the legend. The OR paths are indicated by the dashed lines labeled OR-1, OR-2, and OR-3 in the legend.

Three target locations are presented with both the ML model (in solid lines) and the OR model (in dashed lines), all originating from the same starting location. The OR paths will initially circumnavigate CONUS, where no threats exist, to a location much closer to the intended target. The ML model will (at least as currently trained) weave through the threats as best as possible in a generally straight line. The ML model potentially performs better along the Eastern Seaboard because its path is not constrained to the edges of the map's discretization as described previously. This suggests that the ML model can weave and thread the needle better than the OR model in some situations. Ultimately, parity is expected between these models for locations that are precisely on the East Coast, since no threats exist between the starting location and the East Coast.

Observing all the paths taken by the two different models toward the target locations is informative as to how the models differ in their strategy. Figure 2.9 shows all possible paths.

Figure 2.9. All 1,376 Paths to the Targets for the Operations Research Model and the Machine Learning Model

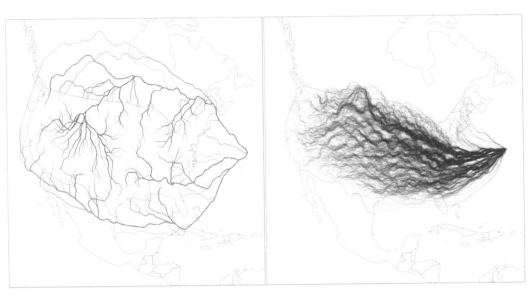

NOTE: The image on the left depicts all possible paths that the OR approach found. The image on the right depicts all possible paths that the ML approach found.

The OR model on the left in Figure 2.9 takes dissimilar trajectories toward the targets. Thus, two targets that are close to each other in terms of distance could result in substantially different path lengths for ingress. On the other hand, the ML model, shown on the right in Figure 2.9, has more direct trajectories toward the targets, resulting in shorter paths overall but at higher risk, as previously established. These results are reminiscent of AlphaStar: The advantage of speed is maintained, and, instead of complex strategies and responses, a brute-force approach, with superior skill at local engagement, is employed.

In our tests, the OR approach almost always found a more-efficient solution, as defined by our utility function, than the ML approach. However, ML has two advantages. First, ML came up with solutions much faster.[16] To generate a new path in response to an unexpected threat on the hardware we used, the OR approach would take many seconds to minutes depending on the granularity of the map discretization.[17] The ML approach generated a new path from the visible world at every step, taking just milliseconds to do so.

Second, the OR approach would often find extremely precise and careful paths, while the ML approach would wobble. This led to the ML approach generating more robust routes with higher tolerance for uncertainties. For example, wind in the real world can cause the aircraft to crab (set the angle of the aircraft differently from the direction of net motion over ground), and the ML approach

---

[16] The ML approach is faster than the OR approach, but there are caveats. The runtime for the OR approach is directly affected by the quantization of the simulation environment; it typically uses 5–30 km granularity. The granularity, or number of discrete points that the OR algorithm can consider, could be decreased to the point where the OR algorithm can generate a path as fast as the ML model. However, the loss in precision would render very bad results (uniformly worse than those generated by the ML model). Increasing the discretization raises both path precision and OR runtime. Setting a fair level of granularity for speed comparisons is not possible because the ML approach operates on continuous space.

[17] The hardware we used was a commercial laptop with an integrated Intel UHD Graphics 620 (GT2).

can tolerate this shift. Note that we trained the RL algorithm to respond to pop-up threats (when a previously unknown early warning radar suddenly appears in front of the aircraft) and uncertain radar locations (in each simulation, radar locations are generated at slightly different locations). However, for a fair apples-to-apples path comparison, we turned the pop-up threats off to generate Figure 2.9.

We consider relative experiences to be of critical importance. Although AI is currently inferior, understanding its relative strengths and weaknesses will enable better AI application as it progresses toward superhuman capability. For example, it is difficult to train an AI model (and it takes a long time), but the benefit is fast (real-time, in our case) route suggestions. Understanding which aspects of AI can be better enables better planning. This assumes a constant rate of improvement across skills, which seems plausible. In our experiments, we found that

- on average, the ML model spent 2.5 times more time within the radius of enemy assets compared with the OR approach
- the ML model was able to reach the target in four-fifths the time of the OR algorithm
- an increase in the density of threats and the complexity of the overall environment made the ML approach more appealing.

The OR algorithm provides a single analytic solution in the form of an expected value of detection time and shot opportunities; there is no variance in expected value from run to run, and the outcome was reasonably normal. On the other hand, the ML model would vary significantly: A trained RL can be stochastic in nature, so for our analysis, we generated ten paths and selected the best one. When the time needed for the RL to calculate a path is milliseconds, there is a great deal of opportunity to find optima depending on real-world timelines. Table 2.1 compares path attributes for the OR and ML approaches.

Table 2.1. Operations Research and Machine Learning Outcomes

	OR	ML
Enemy detection	Average risk level: 1.12	Average risk level: 2.91
Path length	Average: 5,070 km	Average: 3,938 km
Cost to develop	Low	High
Response time	Seconds to minutes	Milliseconds
Predictability	Fixed given a utility function and situation	Varies with each run

NOTE: Enemy detection is the sum of the time spent within the range of enemy detection capabilities. Results are shown across successful missions (reaching the target).

Training AI models has become easier and more straightforward. For example, AutoML scripts can automate some steps, such as algorithm selection and hyperparameter optimization. However, many steps in the development process still require a human touch (e.g., aspects of data cleaning, feature selection, and designing network architectures can be manual). During experimentation, several attempts at model training were scratched because the reward function was slightly miscalibrated. Once the punishment for detection was too high, the model preferred to avoid the target and crash into the ocean rather than going through the adversary's air defense systems.

ML models in highly complex domains, such as protein folding or text generation, typically improve with increases in model size and training iterations. However, mission planning is computationally relatively simple. Once a model delivered a *good* performance (informally defined as a human not seeing obvious improvements), more training did not produce flight paths that recorded a substantially lower threat from enemy radar, lower flight times, or improvements on any other metric. We also frequently encountered local maxima that were very far from global maxima, which required human intervention. During hyperparameter optimization, we found that the best-performing model was a relatively simple two-layer model. While human expertise was critical, we did not find that the resulting model needed to be particularly complex.

From our work with RTAM, we take away the following key points:

- The ML model usually underperformed when compared with our OR approach.
- The ML model outperformed in one-sixth of cases and had a substantial speed advantage.
- The flexibility to incorporate multiple methods and approaches in mission planning is useful, even if one method or approach is usually better than the other.
- In simple representations, the expense of training for a very long period is unnecessary.
- No amount of computing power, time, or money is an adequate substitute for expertise.

Chapter 3

Making a Reinforcement Learning Environment Out of the Advanced Framework for Simulation, Integration, and Modeling

The RTAM toy problem discussed in Chapter 2 demonstrated key differences between ML- and OR-based approaches toward route planning. In this chapter, we discuss the development of tools to enable ML-based approaches for a simulation tool used by the DAF. The DAF uses AFSIM, a C++-based multidomain modeling and simulation framework, to conduct analysis to enable wargaming, experimentation, and mission planning. Shared by the Air Force Research Laboratory (AFRL) with partners under an Information Transfer Agreement, AFSIM is licensed to more than 275 government, industry, and academic organizations.[18] Figure 3.1 provides a publicly releasable screenshot of AFSIM.

---

[18] Jaclyn Knapp, "Information Transfer Agreement Enables AFRL Software Sharing with Industry," Wright-Patterson Air Force Base, March 10, 2017; Timothy D. West and Brian Birkmire, "AFSIM: The Air Force Research Laboratory's Approach to Making M&S Ubiquitous in the Weapon System Concept Development Process," *Journal of the Cyber Security & Information Systems Information Analysis Center*, Vol. 7, No. 3, Winter 2020.

Figure 3.1. Advanced Framework for Simulation, Integration, and Modeling Screenshot

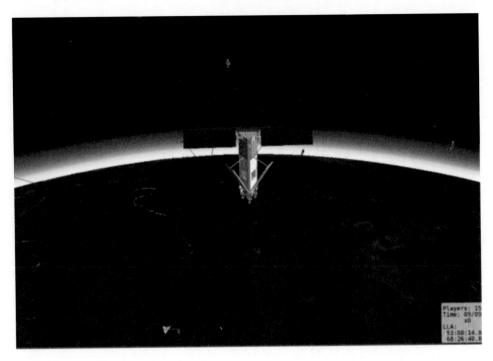

SOURCE: Wright-Patterson Air Force Base, "Advanced Framework for Simulation, Integration and Modeling Software," webpage, U.S. Air Force, undated.

Grey literature sources and experts we interviewed agreed that a substantial contributor to AFSIM's popularity is the degree of open access.[19] Not only is it freely available to partners with an International Traffic in Arms Regulations–compliant environment, but the underlying source code is included. The design is heavily modular, so a controlled unclassified information version can be distributed with relatively minimal information about weapon systems, and modules with more details can be distributed at other classification levels. AFSIM also supports a custom domain-specific language, AFSIMScript, which has been praised as more accessible and which we found helpful in our own work.

The modularity of AFSIM enables work like ours, which can easily be applied to a user's existing system without any additional modification. However, mere modularity is not enough to solve all problems. Originally developed by Boeing in 2003, the code that would become AFSIM was based on C++ because it was widely used at the time and would have provided speed benefits that were more critical in an era when central processing units were orders of magnitude slower than they are in 2023. However, C++ declined in popularity over the intervening two decades, and the language is not a native host of any current popular ML frameworks. The language of choice for most modelers and researchers in industry today is Python, particularly in cases for which performance sensitivity is useful but not a high priority.

In RL parlance, a *gym* is an environment for developing agents. It offers an action and observation space that enables a given policy to be evaluated rapidly. Speed is critical for gyms because RL relies on

---

[19] West and Birkmire, 2020.

iterating models a large number of times. Gyms are necessary for efficient model training, testing, and refinement because interacting with the real world is much more expensive and dramatically slower.

From 2019 to 2020, RAND researchers developed a prototype RL gym with a simplified radar model and a significantly simplified feature set compared with AFSIM called AFGYM.[20] By limiting its scope, RAND researchers were able to demonstrate effective learning and show that the increased performance resulted in far less RL training time. AFGYM has generated much interest from industry and government, indicating a capability gap and appetite for RL modeling infrastructure across the U.S. Department of Defense. However, AFGYM's lack of features has meant there has been minimal use by the broader community.

We produced an end-to-end integration that enabled RL training within AFSIM using the Python TensorFlow framework. We demonstrated this RL framework with two proof-of-concept scenarios: Using virtual aircraft inputs, we showed that it is possible to follow a leader aircraft and to perform automated sensing of objects within an area of operations. The integration with Python, the lingua franca of AI, opens up AFSIM RL exploration to ML researchers who might not be familiar with AFSIM code.

This was not just the first freely available gym that worked within AFSIM, but it was also the first one anywhere. Solving end-to-end integration required that someone understand the domain and the needs of warfighters and planners and be an experienced ML practitioner. Prior RAND researcher work in this area was good as an ML demonstration, but it did not result in a useful environment for others that met all needs. By implementing a much closer match, the solution from RAND researchers was something that could be widely used.

However, even though Python is popular, it is not perfect. Most critically, Python is substantially slower to run, particularly at set-up time (this is partially because Python's type system forces a lot of work to be done by the interpreter at runtime, whereas C++ is compiled in advance). Our version, critically, does not require a Python harness after training, which would cause substantial delays to runtime and is awkward for running with other modeling infrastructure; we operate under the premise that Python is valuable for training, but removing that requirement when running the model for the rest of the project (for which the AI has been trained) carries intrinsic value. Trained models can be deployed directly to AFSIM. It is possible to run a model without involving any Python, which supports users who only know C++. AFSIM was built in C++ and supports C++ commands, in addition to AFSIMScript, and TensorFlow Lite exports to C++.

The gym works for any environment that AFSIM can model, including air, land, sea, cyber, and space. Similarly, while we looked at a model with a very high degree of granularity, the TensorFlow setup would work equally well at the campaign level. Using this gym, we prototyped several RL solutions within AFSIM: keeping an inverted pendulum upright, follow-the-leader, and autonomous sensing.

Our code was shared with the AFRL by posting it to the AFSIM portal. The AFSIM portal has been used by a diverse number of users, including within the AFRL, the broader DAF community, and users outside the DAF. Some of the people we met with to learn about how the portal is being

---

[20] Li Ang Zhang, Jia Xu, Dara Gold, Jeff Hagan, Ajay K. Kochar, Andrew J. Lohn, and Osonde A. Osoba, *Air Dominance Through Machine Learning: A Preliminary Exploration of Artificial Intelligence–Assisted Mission Planning*, RAND Corporation, RR-4311-RC, 2020.

used included students, industry partners, and AI adoption and implementation groups. We engaged with users through email, one-on-one meetings, presentations, and other methods. Unfortunately, the full scope of users is unknown because the portal does not have a way to track users.

We took away four lessons from this work. First, the modeling and simulation community is excited by and interested in adopting ML tools. Second, current support options to connect open-source tools to Department of Defense systems to produce useful products is limited. Although open access is a critical step, without which this work cannot be done, it is not the final step. Third, taking advantage of existing resources saved substantial time and effort and enabled us to effectively reach a much larger community that already had experience with AFSIM. Finally, there are substantial benefits to updating libraries, frameworks, and even languages to match the needs of the modern practitioner community.

Chapter 4

Findings and Recommendations

Findings

AlphaStar illustrates the difficulty of leveraging RL to solve complex planning problems. Teaching a machine to make strategy-level decisions is challenging because it is difficult to attribute individual actions to strategic goals with long-term reward payoffs. This limitation led us to explore RL in route planning, a more narrowly scoped problem. Although we were successful, the training process involved trial and error, the development of heuristics, and major revisions of the reward function to balance the two objectives (reaching the target and minimizing radar detection risk). For a relatively simple task (pathfinding), we required significant code development (tooling to enable AI), research and development, and computing resources. A cost-benefit analysis should precede AI investment in mission planning and RL.

Compared with OR approaches, AI typically performs worse. Given that OR involves solving well-posed optimization problems, this result is not surprising. However, AI can be more robust and responsive to a changing environment because an OR solution is only meant to solve a static problem.

Despite these caveats, AI is capable of helping out in some planning roles, and using AI in this way will build capacity, experience, and user trust for future AI use. Mission route planning is one example of a narrow AI application that is particularly useful for dynamic threat environments, in which mission packages enter into a complicated air defense environment with pop-up threats.

AI for mission planning requires the development of infrastructure that efficiently connects a simulation environment with an AI framework, which is often written in a different coding language (e.g., TensorFlow or PyTorch). Fortunately, this is a one-time investment for each simulation environment. The DAF should consider such investments and release the infrastructure to government and partner organizations. Developing AI on a common infrastructure is cheaper than having vendors compete on duplicate functionality and enables easier apples-to-apples comparisons between competing RL models.

Implementing AI in mission planning, and in warfare more broadly, is not just a matter of creating a standalone program. **It is critical to support connections to other tools and continuously update those connections as new tools are invented.** Without this ongoing support and effort, the actual use of AI will inevitably lag compared with near-peer adversaries. The DAF will better serve warfighter needs by thinking consciously about future applications and software.

Recommendations

As a first step to apply AI for mission route planning, we recommend scoping out a narrow problem for AI and then prototyping a series of reward functions to encourage learning behaviors one at a time. OR approaches are objectively better but might not be robust or possible in all domains. For very complex decisionmaking over many time steps, it can be impossible to formulate such challenges as optimization-based problems (or mathematically infeasible) and ML approaches are attractive options. Narrowly scoping the problem and iteratively developing the reward function enables researchers to develop complex behavior over time.

We recommend the DAF apply RL mission planning to dynamic route planning for uncrewed systems that are reviewed and judged by operators. For now, the best use of RL in mission planning is as a fast-reacting management system that responds dynamically to threats. This applies both to an on-board drone and to a headquarters that can give an updated flight plan in a matter of seconds rather than minutes. We believe that this approach is attractive for expanding the capabilities of uncrewed systems and that implementation will build user trust and experience. Even when RL provides suboptimal plans, it can suggest immediate actions; operators can take the time gained from a rapid response to develop better plans using their current standard and preferred methods.

Because existing human planning processes make effective use of available information, improvements in the efficiency of outcomes from implementing AI mission planning are relatively small. However, scaling up planning for high-end conflicts for hundreds of daily aircraft sorties creates an untenable demand for human planners within the AOC. At present, **we expect the cost of AI-based mission planning to substantially exceed the benefits if it is envisioned as a total replacement for human planning.** However, the DAF can take steps to lay a foundation for AI use, as recommended previously by RAND researchers.[21]

The DAF should train AI specialists who also have a deep understanding of military mission planning. RL is a difficult research area that relies on experience and heuristics. The research is further complicated by the need for application-specific knowledge. Those who lack familiarity in the area might not recognize undesirable states and behaviors, preventing them from crafting the suitable reward function.

The DAF must prioritize tools and software, not just by creating them but by enabling those resources to be extensible and connectable to existing systems. Existing simulation tools should be extended to be compatible with AI frameworks. Training RL requires putting the algorithm through a wide variety of situations. Feeding in data, targets, and other information automatically will greatly speed up the workflow in training and operational deployment. The strong demand for our code demonstrates that demand exists for tools that are not just compatible with AI but actively integrate it. We have released our infrastructure on the AFRL Gitlab portal, which alleviates the resource burden for future researchers to apply RL with AFSIM.

The DAF should continually monitor the AI RL landscape; paradigm shifts have happened before in RL and can happen again. Although AI advancements occur rapidly in commercial and research communities, the DAF will need to remain vigilant about looking for opportunities to integrate new advancements.

---

[21] Sherrill Lingel, Jeff Hagen, Eric Hastings, Mary Lee, Matthew Sargent, Matthew Walsh, Li Ang Zhang, and David Blancett, *Joint All-Domain Command and Control for Modern Warfare: An Analytic Framework for Identifying and Developing Artificial Intelligence Applications*, RAND Corporation, RR-4408/1-AF, 2020.

Detailed AlphaStar Discussion

Mission planning is the simultaneous assignment of diverse assets, crewed and uncrewed, to accomplish objectives within a limited time frame. For our analysis, we primarily looked at route planning: Once an asset or package has been assigned a particular target, how does it get there and back again?

There was little information available about the commercial use of route-planning software. However, there was considerable public attention, resulting in a wide variety of sources, on AI for computer games that seemed like a reasonable approximation.

Prior competitive AI systems, including video game AI in general, relied heavily on hand-coded elements and decision tree-based heuristics.[22] To create AlphaStar, DeepMind formed a league of ML models that competed against each other. The tactics employed by the league changed over time before converging on the relatively stable strategy that AlphaStar ultimately employed. This was a very expensive process: The compute resources alone would cost more than $12 million to replicate.[23]

An important point to consider is that StarCraft II is a game of imperfect information. Unlike go, chess, and shogi, a fog of war is present that obscures enemy activity. Visibility is limited to areas around a player's units and buildings, which makes scouting to determine an enemy's location, force composition, and future intent critical. Imperfect information forces players to predict what the enemy is building rather than simply maximizing economic or army production efficiency. Players must attack when an enemy will be at its weakest and keep their units safe when the enemy will be at its strongest. Late-game units are more powerful and units from at least the mid-game are required to effectively destroy an enemy base, but early effective attacks can still be valuable if uncontested. Attacking vulnerable workers early can cripple enemy resource production to create a late-game advantage. Figuring out when an enemy will be weak or strong on the basis of limited information is a process that has substantial similarities to the mission planning process during wartime. However, this was not an area in which AlphaStar did well.

Expectations in advance of league matches were mixed. Caster Falcon Paladin stated that although AI was clearly capable of tactical play and decisionmaking, as evidenced by past AI success at chess and go, AI was expected to struggle with such aspects as positioning, micro, and "larger big-picture complex decisionmaking."[24] *Micro* is a term of art used among esports players that refers to high-speed and precise actions, particularly within a single combat, as opposed to *macro*, which refers to broader

---

[22] One prominent StarCraft II example involves programming a swarm of flying units with superhuman multitasking precision. However, as typical of most game-playing AI, this was a hardcoded behavior.

[23] Ken Wang, "DeepMind Achieved StarCraft II Grandmaster Level, but at What Cost?" Medium, January 4, 2020.

[24] A caster is a commentator who provides a spoken overview to help viewers parse what is happening in a videogame. Falcon Paladin, "AlphaStar (P) v MaNa (P) 5-Game Series!—StarCraft2—Legacy of the Void 2018," video, YouTube, January 27, 2019.

strategy across the entire game. As mentioned earlier, AlphaStar took advantage of micro and readily defeated humans battle after battle with stalker units.

AlphaStar exhibited some mimicry of human decisionmaking, which was odd. For example, humans remove rocks from the game because their presence causes misclicks (e.g., attacking the rock instead of the enemy player). However, AlphaStar would remove rocks even though misclicks are not a concern for it. This sort of mistake is not a necessary feature of DeepMind's modeling (AlphaZero learned DeepMind's go model only by playing itself) but is a likely artifact of the training strategy that included human-level players.

One novel tactic known as three oracles, which gained rapid popularity among pros, gave AlphaZero trouble, plausibly because it had never trained against the tactic.[25] While ML models are capable of rapidly responding to new information, they struggle to respond to entirely new strategies that require a paradigm shift. Poor responses to out-of-sample data are a known and continuing problem in ML. Strategic failures were also persistent: "They will construct buildings that block them into their own base, crowd their units into a dangerous bottleneck to get to a cleverly-placed enemy unit, and fail to change tactics when their current strategy is not working."[26] Some strategies enabled by superior micro were considered viable by serious players and adopted by humans.[27]

---

[25] *Three oracles* refers to three specific units created at a particular point in time. LowkoTV, 2019, 5:00.

[26] Rick Korzekwa, "The Unexpected Difficulty of Comparing AlphaStar to Humans," *AI Impacts* blog, September 17, 2019.

[27] LowkoTV, 2019, 10:30–11:03.

Abbreviations

AFRL	Air Force Research Laboratory
AFSIM	Advanced Framework for Simulation, Integration, and Modeling
AI	artificial intelligence
AOC	Air Operations Center
CONUS	continental United States
DAF	Department of the Air Force
ML	machine learning
OR	operations research
RL	reinforcement learning
RTAM	RAND Target Accessibility Model

References

Falcon Paladin, "Alphastar (P) v MaNa (P) 5-Game Series!—StarCraft2—Legacy of the Void 2018," video, YouTube, January 27, 2019. As of September 24, 2022:
https://youtu.be/sB7unYvSKk8?si=FfillIBU8QCejKMt

Fraade-Blanar, Laura, and Brian A. Jackson, "Developing a Winning Safety Strategy for Automated Vehicles," *RAND Blog*, February 18, 2022. As of September 24, 2022:
https://www.rand.org/blog/2022/02/developing-a-winning-safety-strategy-for-automated.html

Geist, Edward, Aaron B. Frank, and Lance Menthe, *Understanding the Limits of Artificial Intelligence for Warfighters:* Vol. 4, *Wargames*, RR-A1722-4, 2024.

Kaplan, Jared, Sam McCandlish, Tom Henighan, Tom B. Brown, Benjamin Chess, Rewon Child, Scott Gray, Alec Radford, Jeffrey Wu, and Dario Amodei, "Scaling Laws for Neural Language Models," arXiv, January 23, 2020.

Knapp, Jaclyn, "Information Transfer Agreement Enables AFRL Software Sharing with Industry," Wright-Patterson Air Force Base, March 10, 2017.

Korzekwa, Rick, "The Unexpected Difficulty of Comparing AlphaStar to Humans," *AI Impacts* blog, September 17, 2019.

Lingel, Sherrill, Edward Geist, Thomas Hamilton, Daniel M. Norton, and Colby P. Steiner, *(U) Toward Continuous Planning for Modern Warfare: A Warfighting-Focused Framework for Operational Planning of Science and Technology Pursuits*, RAND Corporation, RR-A953-1, 2023, Not available to the general public.

Lingel, Sherrill, Jeff Hagen, Eric Hastings, Mary Lee, Matthew Sargent, Matthew Walsh, Li Ang Zhang, and David Blancett, *Joint All-Domain Command and Control for Modern Warfare: An Analytic Framework for Identifying and Developing Artificial Intelligence Applications*, RAND Corporation, RR-4408/1-AF, 2020. As of October 11, 2023:
https://www.rand.org/pubs/research_reports/RR4408z1.html

LowkoTV, "StarCraft 2: AlphaStar (Artificial Intelligence) vs Grand Master League!" video, YouTube, November 12, 2019. As of September 24, 2022:
https://www.youtube.com/watch?v=VsJpWZj9jl0

Menthe, Lance, Li Ang Zhang, Edward Geist, Joshua Steier, Aaron B. Frank, Eric Van Hegewald, Gary J. Briggs, Keller Scholl, Yusuf Ashpari, and Anthony Jacques, *Understanding the Limits of Artificial Intelligence for Warfighters:* Vol. 1, *Summary*, RR-A1722-1, 2024.

Schulman, John, Filip Wolski, Prafulla Dhariwal, Alec Radford, and Oleg Klimov, "Proximal Policy Optimization Algorithms," arXiv, August 28, 2017.

Silver, David, Thomas Hubert, Julian Schrittwieser, Ioannis Antonoglou, Matthew Lai, Arthur Guez, Marc Lanctot, Laurent Sifre, Dharshan Kumaran, Thore Graepel, Timothy Lillicrap, Karen Simonyan, and Demis Hassabis, "A General Reinforcement Learning Algorithm That Masters Chess, Shogi, and Go Through Self-Play," *Science*, Vol. 362, No. 6419, December 2018.

Steier, Joshua, Erik Van Hegewald, Anthony Jacques, Gavin S. Hartnett, and Lance Menthe, *Understanding the Limits of Artificial Intelligence for Warfighters:* Vol. 2, *Distributional Shift in Cybersecurity Datasets*, RAND Corporation, RR-A1722-2, 2024.

Thaler, David E., and David A. Shlapak, "Perspectives on Theater Air Campaign Planning," RAND Corporation, MR-515-AF, 1995. As of October 11, 2023:
https://www.rand.org/pubs/monograph_reports/MR515.html

Vinyals, Oriol, Igor Babuschkin, Wojciech M. Czarnecki, Michaël Mathieu, Andrew Dudzik, Junyoung Chung, David H. Choi, Richard Powell, Timo Ewalds, Petko Georgiev, Junhyuk Oh, Dan Horgan, Manuel Kroiss, Ivo Danihelka, Aja Huang, Laurent Sifre, Trevor Cai, John P. Agapiou, Max Jaderberg, Alexander S. Vezhnevets, Rémi Leblond, Tobias Pohlen, Valentin Dalibard, David Budden, Yury Sulsky, James Molloy, Tom L. Paine, Caglar Gulcehre, Ziyu Wang, Tobias Pfaff, Yuhuai Wu, Roman Ring, Dani Yogatama, Dario Wünsch, Katrina McKinney, Oliver Smith, Tom Schaul, Timothy Lillicrap, Koray Kavukcuoglu, Demis Hassabis, Chris Apps, and David Silver, "Grandmaster Level in StarCraft II Using Multi-Agent Reinforcement Learning," *Nature*, Vol. 575, November 2019.

Wang, Ken, "DeepMind Achieved StarCraft II GrandMaster Level, but at What Cost?" Medium, January 4, 2020.

Waters, K. Houston, "Hanscom AFB Team Supports JADC2 Through Agile Software Development," U.S. Air Force, June 7, 2021.

West, Timothy D., and Brian Birkmire, "AFSIM: The Air Force Research Laboratory's Approach to Making M&S Ubiquitous in the Weapon System Concept Development Process," *Journal of the Cyber Security & Information Systems Information Analysis Center*, Vol. 7, No. 3, Winter 2020.

Wright-Patterson Air Force Base, "Advanced Framework for Simulation, Integration and Modeling Software," webpage, U.S. Air Force, undated. As of September 24, 2022:
https://www.wpafb.af.mil/News/Art/igphoto/2001709929/

Zhang, Li Ang, Yusuf Ashpari, and Anthony Jacques, *Understanding the Limits of Artificial Intelligence for Warfighters:* Vol. 3, *Predictive Maintenance*, RAND Corporation, RR-A1722-3, 2024.

Zhang, Li Ang, Jia Xu, Dara Gold, Jeff Hagen, Ajay K. Kochhar, Andrew J. Lohn, and Osonde A. Osoba, *Air Dominance Through Machine Learning: A Preliminary Exploration of Artificial Intelligence–Assisted Mission Planning*, RAND Corporation, RR-4311-RC, 2020. As of October 11, 2023:
https://www.rand.org/pubs/research_reports/RR4311.html